NANDINI TEWARI'S
BEYOND THE KUIPER BELT

ILLUSTRATED BY
AISHANI M. NANDA

Made with ❤ on the Notion Press Platform
www.notionpress.com

BEYOND THE KUIPER BELT

Nandini Tewari, a 17-year-old aspiring astrophysicist from New Delhi, India, is an avid reader and writer with a passion for physics and mathematics. A multi-talented individual, she excels in research, chess, and kathak dance, and is also a skilled debater with a degree in scuba diving. Nandini's diverse interests and talents reflect her dynamic approach to learning and discovery.

FOR BABA,
FOREVER AND ALWAYS

CONTENTS

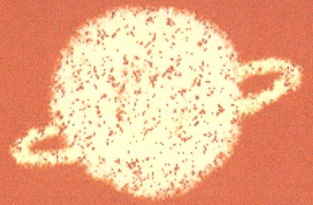

PREFACE

BEYOND THE KUIPER BELT EMBARKS ON AN ODYSSEY INTO THE COSMOS, INTERTWINING THE MARVELS OF SPACE EXPLORATION WITH THE INNATE CURIOSITY OF YOUNG MINDS. AS SOMEONE WHO HAS LONG BEEN CAPTIVATED BY BOTH THE MYSTERIES OF THE UNIVERSE AND THE ART OF STORYTELLING, MY INTENT IN WRITING THIS BOOK IS TO IGNITE A SENSE OF WONDER WHILE MAKING INTRICATE SCIENTIFIC CONCEPTS BOTH ACCESSIBLE AND ENGAGING. DRAWING UPON MY OWN PERSONAL EXPERIENCES AND PASSIONS —FROM THE STRATEGIC INTRICACIES OF CHESS TO THE EXPRESSIVE GRACE OF KATHAK DANCE AND THE SERENE DEPTHS ENCOUNTERED IN SCUBA DIVING—I HAVE FOUND STRIKING PARALLELS BETWEEN OUR EVERYDAY LIVES AND THE VAST, ENIGMATIC UNIVERSE THAT STRETCHES FAR BEYOND OUR REACH.

IN THIS BOOK, WE WILL JOURNEY TOGETHER THROUGH SPACE, EXPLORING PLANETS, STARS, BLACK HOLES, AND THE WONDERS THAT LIE BEYOND THE KUIPER BELT. MY GOAL IS TO DEMYSTIFY THE COMPLEXITIES OF ASTROPHYSICS BY WEAVING TECHNICAL CONCEPTS INTO NARRATIVES THAT ARE APPROACHABLE, USING ANALOGIES GROUNDED IN FAMILIAR EXPERIENCES TO DEMONSTRATE THAT SCIENCE NEED NOT REMAIN CONFINED TO ACADEMIA—IT IS FOR EVERYONE. WHETHER YOU ARE AN ASPIRING ASTROPHYSICIST LIKE MYSELF OR SIMPLY SOMEONE INTRIGUED BY THE UNKNOWN THAT SPARKLES IN THE NIGHT SKY, I HOPE THIS BOOK SERVES AS A GATEWAY TO YOUR OWN EXPLORATION OF THE COSMOS.

I INVITE YOU TO VENTURE BEYOND THE FAMILIAR AND DISCOVER THE BOUNDLESS MYSTERIES THAT AWAIT BEYOND THE KUIPER BELT.

THE BIG BANG

97 YEARS AGO, WE DIDN'T HAVE TELEVISIONS, 174 YEARS AGO, WE DIDN'T HAVE PHONES, 1,500 YEARS AGO, THE GAME OF CHESS DIDN'T EXIST AND 14 BILLION YEARS AGO, THERE WAS NOTHING: NO GALAXIES, NO CARS, NO KNIGHTS AND NO LIGHT. IT WAS EMPTY. BARREN. THEN IT ALL STARTED WITH A DENSE AND HOT BUBBLE. THE WORLD, AS WE KNOW IT TODAY, FIT INTO THAT ONE PINHEAD MIXED WITH LIGHT AND ENERGY.

AND THEN THE BUBBLE BURST.

AN EXPLOSION TOOK PLACE AND THE UNIVERSE WAS BORN. IN A FRACTION OF A SECOND, IT EXPANDED AT A BRILLIANT RATE.

THE PARTICLES GROUPED TOGETHER TO FORM PROTONS AND NEUTRONS, AND SO WERE BIRTHED THE FIRST TWO GASSES: HYDROGEN AND HELIUM. AS THE TEMPERATURE COOLED DOWN EVEN FURTHER, THE NUCLEI OF ATOMS COULD NOW FORM BONDS WITH ELECTRONS, AND SO WAS THE FIRST ATOM FORMED! SUBSEQUENTLY, STARS, PLANETS AND GALAXIES WERE BROUGHT INTO EXISTENCE, AND THE UNIVERSE WAS FINALLY COMPLETE.

BUT WAS THIS IT?

THE UNIVERSE, AS DEFINED BY GEORGES LEMAÎTRE AND EDWIN HUBBLE, IS STILL EXPANDING! TO IMAGINE THAT ALL OF THIS STARTED FROM A TINY, TINY DOT, TO AN ENTITY THAT IS STILL INFLATING IS FASCINATING!

THE SOLAR SYSTEM

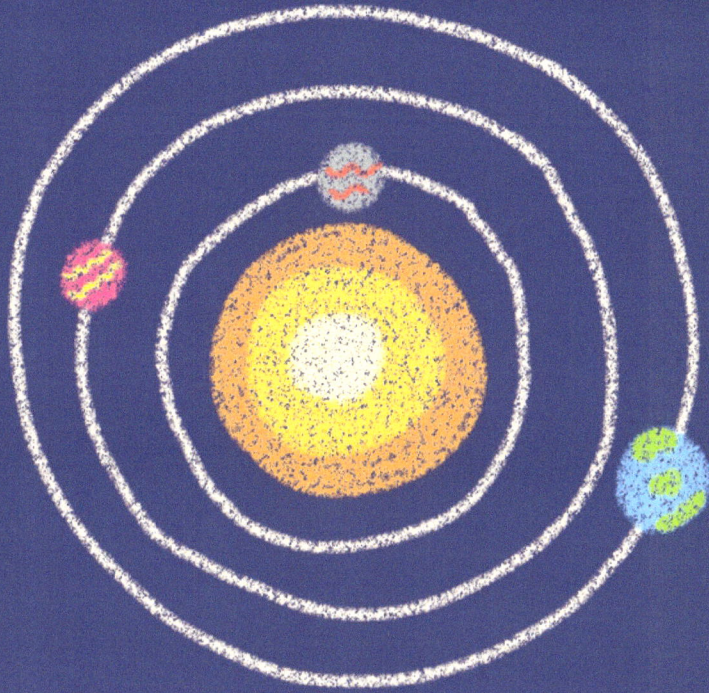

THE SOLAR SYSTEM, AS THE NAME SUGGESTS, IS A PLANETARY SYSTEM CENTRED AROUND OUR LOCAL STAR, THE SUN. THE WORD "SOLAR" IS DERIVED FROM THE LATIN WORD FOR THE SUN, "SOLIS". OUR SOLAR SYSTEM IS MADE UP OF 8 PLANETS ORBITING THE SUN, THEIR MOONS, DWARF PLANETS, AND A PLETHORA OF ASTEROIDS, COMETS AND OTHER CELESTIAL BODIES.

THE PLANETARY STRUCTURE OF THE SOLAR SYSTEM IS AS FOLLOWS:

MERCURY, VENUS, EARTH, MARS, JUPITER, SATURN, URANUS AND NEPTUNE, WITH MERCURY BEING THE CLOSEST PLANET TO THE SUN AND NEPTUNE BEING THE FARTHEST.

THE PLANETS, BASED ON THEIR COMPOSITION, HAVE BEEN DIVIDED INTO TWO MAIN GROUPS: ROCKY PLANETS AND GAS GIANTS. THE ROCKY PLANETS, NAMELY MERCURY, VENUS, EARTH AND MARS, ARE DIVIDED FROM THE GAS GIANTS, NAMELY, JUPITER, SATURN, URANUS AND NEPTUNE BY AN ASTEROID BELT.

WE WILL DIVE DEEPER INTO THE WORLD OF OUR PLANETS FURTHER IN THE BOOK.

THE SUN

ALL OF US KNOW WHAT THE SUN IS! THE FIRST THING YOU SEE IN THE MORNING, THE REASON BEHIND OUR ICE CREAM CRAVINGS, THE INSTIGATOR OF OUR SUNGLASSES, IT IS ALL THE SUN'S DOING!

BUT THE SUN ISN'T AS CLOSE AS IT SEEMS. THE DISTANCE BETWEEN THE EARTH AND THE SUN IS 93 MILLION MILES! NOW THAT WOULD BE A TIRING CAR RIDE! THE SUN IS A HOT, DENSE AND BRIGHT BALL OF GASES, LIGHT AND ENERGY.

THE SUN PLAYS AN EXTREMELY CRUCIAL ROLE IN MAINTAINING LIFE ON EARTH. WITHOUT THE SUN, WE WON'T HAVE LIGHT ON OUR PLANET, NO DAYS AND NO NIGHTS. WITHOUT ITS WARMTH, WE WOULD ALL FREEZE, AND PLANTS AND ANIMALS WOULD SHIVER TO DEATH. THEREFORE, THE SUN IS LIKE THE PRESENCE OF OUR BEST FRIEND IN SCHOOL, WITHOUT WHOM WE SIMPLY CANNOT EXIST!

THE SUN IS PLACED AT THE VERY CENTRE OF THE SOLAR SYSTEM AND IS BY FAR THE LARGEST OBJECT IN THE SOLAR SYSTEM, WITH A DIAMETER OF AROUND 865,000 MILES AND A MASS OF AROUND $1.9891*10^{30}$ KG!

THE CORE COMPONENTS OF THE SUN ARE SURPRISINGLY, TWO GASES: HYDROGEN AND HELIUM! THE HYDROGEN IS CRAMMED IN THE CORE OF THE SUN, VERY MUCH LIKE OUR HOMEWORK AT THE BOTTOM OF OUR BACKPACKS. THE IMMENSE AMOUNT OF PRESSURE APPLIED ON THE HYDROGEN CHANGES IT TO HELIUM THE SURFACE TEMPERATURE OF THE SUN IS 10,000 DEGREES FAHRENHEIT!

THE PLANETS

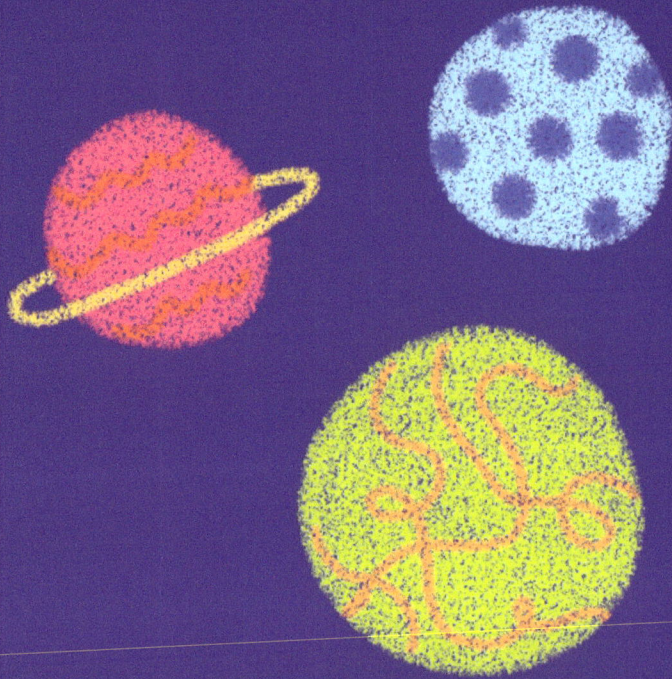

1, 2, 3, 4, 5, 6, 7, 8 STOP. THE NUMBER OF PLANETS IN OUR SOLAR SYSTEM, THE NUMBER OF SQUARES ON EACH SIDE OF THE CHESS BOARD, AND THE NUMBER OF SCOOPS OF ICECREAM I WOULD LIKE TO HAVE, AND THE DAY ON WHICH MY DAD AND I WERE BORN, ALL HAVE ONE THING IN COMMON, THE NUMBER 8.

OUR SOLAR SYSTEM IS A CONGLOMERATE OF 8 DIFFERENT PLANETS, EACH WITH A SET OF UNIQUE CHARACTERISTICS. BEFORE WE DELVE INTO THE INTRICACIES OF EACH PLANET, LET US UNDERSTAND THE BASIC DIVISION OF THESE CELESTIAL BODIES.

IN THE ICECREAM STORE OF THE SOLAR SYSTEM, PLANETS ARE AVAILABLE IN MAINLY TWO FLAVOURS: THE ROCKY PLANETS, WHICH ARE MERCURY, VENUS EARTH AND MARS, AND THE GAS GIANTS, ENTAILING JUPITER, SATURN, URANUS AND NEPTUNE.

NOW LET US DEEP DIVER INTO THE WORLD OF THESE HEAVENLY ORBS!

MERCURY

THE FIRST SCOOP IN LINE, MERCURY, IS THE SMALLEST AND THE CLOSEST PLANET TO THE SUN IN OUR SOLAR SYSTEM. IT WAS NAMED AFTER THE MESSENGER OF THE ROMAN GODS, SINCE IT IS ALSO THE FASTEST PLANET OUT OF THE LOT, COMPLETING ONE ORBIT IN 88 DAYS! THAT IS LIKE HAVING A BIRTHDAY PARTY EVERY 3 MONTHS, HOW I WISH I COULD LIVE ON MERUCRY!

BUT WAIT, DON'T BE TOO QUICK MAKING THAT WISH! DUE TO ITS CLOSENESS TO THE SUN, MERCURY IS ALSO SUPER HOT, MAKING HUMAN LIFE ON THE PLANET IMPOSSIBLE.

MERCURY'S DARK GRAY SURFACE, FULL OF CRATERS AND DUST DOES NOT HAVE AN ATMOSPHERE OR ANY MOONS. HUMANS HAVE BEEN AWARE OF MERCURY'S EXISTENCE SINCE A LONG TIME, AS IT CAN BE SEEN WITHOUT THE USE OF ADVANCED TELESCOPES OR MACHINERY!

VENUS

NOW COMING TO THE SECOND SCOOP, ALSO THE ONE WHICH MELTS THE FASTEST, LET US SAY HELLO TO THE HOTTEST PLANET OF OUR SOLAR SYSTEM: VENUS! WHILE VENUS ISN'T THE CLOSEST PLANET TO THE SUN, IT IS STILL THE HOTTEST DUE TO THE GREENHOUSE GASES AND THE THICK SULPHRIC CLOUDS MAKING UP ITS ATMOSPHERE.

THE GASES AND CLOUDS TRAP ALL THE HEAT, MAKING THE PLANET SCORCH AND FLAME!

THE TOPOGRAPHY OF VENUS IS FULL OF VOLCANOES AND MOUNTAINS. TERMED AS "EARTH'S TWIN", IT IS APPROXIMATELY THE SAME SIZE AS THE EARTH.

14 THE MOVEMENT OF THE PLANET IS QUITE PECULIAR, WITH A SPIN OPPOSITE TO MOST OF THE OTHER SOLAR SYSTEM PLANETS, AN EXTREMELY SLOW ROTATION RATE OF 243 EARTH DAYS, BUT A REVOLUTION RATE FASTER THAN THAT OF THE EARTH, WHERE A YEAR IS JUST 225 DAYS LONG.

THAT MEANS, A DAY ON VENUS IS LONGER THAN A YEAR ON VENUS! WHAT A WEIRD PLANET!

will melt!

clouds of sulfuric gases

SIMILAR TO MERCURY, VENUS ALSO DOES NOT HAVE ANY MOONS, AND HAS ALSO BEEN KNOWN SINCE THE ANCIENT DAYS, DUE TO ITS PROMINENT AND VISIBLE PRESENCE IN THE NIGHT SKY, GAINING IT NAMES OF THE "MORNING AND EVENING STAR"!

EARTH

AAAAND WE'RE HOME. THIS TERRESTRIAL BLUE AND GREEN GEOID IS ALL THE LIFE WE KNOW OF TILL NOW. THIS PLANET HAS IT ALL: MOUNTAINS AND PLAINS, OCEANS AND RIVERS, COWS AND DOGS, YOU AND I AND A TUB OF ICE CREAM!

THE ATMOSPHERE OF THE EARTH IS EXTREMELY PRECIOUS. IT IS MADE UP PRIMARILY OF NITROGEN AND HAS OXYGEN IN GREAT AMOUNTS TO SUPPORT LIFE. THE ATMOSPHERE ALSO PROTECTS US FROM FOREIGN OBJECTS, SUCH AS METEORS, COMETS AND METEOROIDS, AS IT BREAKS DOWN THESE BODIES BEFORE THEY CAN REACH THE EARTH'S SURFACE.

70% OF OUR EARTH IS MADE UP OF WATER BODIES, AND WE HAVE ONE NATURAL SATELLITE, OUR VERY OWN MOON!

MARS

oreo pieces? yum!

beep boop! I'm a rover

WELL, AFTER VISITING SOME OF THE HOTTEST PLANETS OF OUR SOLAR SYSTEM, LET ME INTRODUCE YOU TO THE RASPBERRY ICECREAM SCOOP OF THE ROCK PLANETS: MARS! THE RUSTY IRON IN THE GROUND MAKES THIS PLANET RED TO THE CORE, HENCE EXPLAINING THE TITLE "THE RED PLANET".

THE OREO PIECES IN THE SCOOP ARE MARKINGS FROM DARK STAND AND DUST SCATTERED BY THE HIGH-FLOWING STREAMS OF CARBON DIOXIDE.

WHILE MARS IS HALF THE SIZE OF THE EARTH, THE ATMOSPHERIC CONDITIONS AND MUCH OF THE TOPOGRAPHY ODDLY RESEMBLE THAT OF EARTH.

WITH DIFFERENT SEASONS, POLAR ICE CAPS, VOLCANOES, CANYONS, AND WEATHERS, MARS IS LITERALLY EARTH'S SECOND SCOOP! THERE IS EVEN A POSSIBILITY OF LIFE ON THE PLANET, AS TRACES OF ANCIENT FLOODS AND LIQUID SALTY WATER PRESENT ON SOME OF THE MARTIAN HILLSIDES HAVE BEEN FOUND, WITH WATER EXISTING ON MARS IN ICY DIRT AND THIN CLOUDS.

ALONG WITH THE RED SCOOP, ONE GETS TWO SLICES OF THE ROUND WHITE OREO CREAM, ALSO KNOWN AS MARS' MOONS, PHOBOS AND DEIMOS.

MOREOVER, THIS PLANET IS THE ONLY ONE IN THE ICE CREAM SHOP WHERE HUMANS HAVE SUCCESSFULLY SENT ROVERS, INSTRUMENTS WHICH MOVE ON THE SURFACE OF THE EARTH, TAKING PICTURES AND COLLECTING SAMPLES OF THE SOIL AND ROCKS.

JUPITER

SAY HELLO TO THE LARGEST CONE REQUIREMENT OF THE LOT, THE LARGEST PLANET OF THE SOLAR SYSTEM: JUPITER!

THIS PLANET HAS CLOUDS OF HYDROGEN AND HELIUM WITH SMALL AMOUNTS OF WATER DROPLETS, ICE CRYSTALS, AMMONIA CRYSTALS AND OTHER ELEMENTS WHICH GIVES IT THE UNIQUE SALTED CARAMEL COLOUR.

THE CARAMEL DRIZZLE STREAKS ON THE PLANET ARE CAUSED BY THE VARIATIONS IN THE CHEMICAL COMPOSITION AND TEMPERATURE OF THE ATMOSPHERIC GASES ON JUPITER.

THE LIGHT-COLOURED BANDS ARE CALLED "ZONES" SIGNIFYING GAS STRIPES WHICH ARE RISING AND THE DARK-COLOURED BANDS CALLED "BELTS" SHOWCASING THE GAS STRIPES WHICH ARE SINKING.

AN INCREDIBLE FEATURE OF THIS FLAVOUR IS THE GREAT RED SPOT, A STORM ON JUPITER WHICH HAS BEEN GOING ON FOR HUNDREDS OF YEARS.

THE SECRET BEHIND THIS STORM'S LONG RUN IS ITS CREATION AND CONTINUATION BY TWO JET STREAMS MOVING IN OPPOSITE DIRECTIONS.

JUPITER BEING THE FIRST IN ROW OF THE GAS GIANTS QUEUE DOES NOT HAVE A SOLID SURFACE LIKE THE ROCKY PLANETS. IT IS MADE UP OF DIFFERENT GASES HELD TOGETHER BY THE EXISTENCE OF A SOLID INNER CORE APPROXIMATELY THE SIZE OF THE EARTH.

MULTIPLE NATURAL AND ARTIFICIAL SATELLITES HAVE ACCOMPANIED THIS PLANET ON ITS CONTINUOUS VOYAGE AROUND THE SUN, WITH 80 CONFIRMED MOONS AND SEVERAL SPACECRAFT ORBITERS AND PROBES, SUCH AS PIONEER 10 AND 11, VOYAGER 1 AND 2, CASSINI, NEW HORIZONS, AND FINALLY, THE ONE WITH AN INTERESTING STORY THAT WE SHALL DISCUSS LATER IN THE WORKBOOK, JUNO.

SATURN

AH, YOU MUST BE TIRED BY NOW, TOURING SO MANY PLANETS, BUT THIS ONE IS SURE TO GET YOU SPINNING AROUND.

I MUST WELCOME YOU TO MY FAVOURITE OF THE BUNCH, THE ONE WITH A RING AND A ROOM FULL OF MOONS, THERE IT IS: SATURN!

THE FRENCH VANILLA OF THE LOT WITH A FEW DRIZZLES OF CHOCOLATE SYRUP (YOU CAN NEVER GO WRONG WITH CHOCOLATE SYRUP) HAS A WAFER RING AROUND IT MADE UP OF DUST, ROCKS, ICE AND THE REMNANTS OF COMETS AND ASTEROIDS WHICH DISINTERGRATE POST COLLIDING WITH THE PLANET.

THE ITALIAN ASTRONOMER, GALILEO GALILEI DISCOVERED THIS PLANET IN THE YEAR 1610.

SIMILAR TO JUPITER, THIS PLANET ALSO BEING A GAS GIANT IS MADE UP OF PRIMARILY HYDROGEN AND HELIUM, ALONG WITH A THICK ATMOSPHERE.

SIMILAR TO CHOCOLATE CHIPS ARE SATURN'S MOONS, WITH A TOTAL OF 145 OF THEM THAT WE KNOW SO FAR, THE MOST NUMBER OF MOONS IN THE ENTIRE SOLAR SYSTEM!

URANUS

THE PLANET WHICH IS THE ULTIMATE RULE BREAKER, ON A QUEST TO BECOME THE MOST UNIQUE FLAVOUR OF ALL: URANUS!

THIS PLANET HAS METHANE ALONG WITH HYDROGEN AND HELIUM IN THE ATMOSPHERE, GIVING IT THE COTTON CANDY BLUE COLOUR!

EVEN THOUGH IT IS COUNTED IN THE GAS GIANTS GROUP, IT ACTUALLY IS AN "ICE GIANT"; DUE TO IT BEING MADE UP OF WATER, AMMONIA FLUIDS, AND OTHER ICY MATERIALS ABOVE A SOLID CORE.

ANOTHER "PINT" OF DIFFERENCE, URANUS ROTATES ON ITS SIDE AND IN THE DIRECTION OPPOSITE TO MOST OF THE PLANETS.

SIMILAR TO SATURN, EVEN URANUS HAS RINGS, BUT THEY ARE FAINTER IN VISIBILITY AND ARE ALSO PERPENDICULAR TO SATURN'S RINGS, OWING TO THE DIRECTION OF ITS ROTATION.

NEPTUNE

"HELLO HELLO, THE CONNECTION MIGHT FAIL BECAUSE OF THE WINDS AND STORMS AT OUR FINAL PLANETARY LOCATION. ANYWAY, THERE IS BARELY ANY LIGHT HERE, NO POINT OF RECORDING THIS."

REPORTING FROM THE FINAL PLANET, THIS IS OUR SITUATION, BECAUSE WE'RE FINALLY AT NEPTUNE! THIS BLUE RASPBERRY SORBET IS EXTREMELY COLD, CLOUDY, WINDY AND DARK, THE DISTANCE FROM THE SUN BEING 30 TIMES THAT OF THE EARTH!

NEPTUNE IS SIMILAR TO URANUS, WITH METHANE AND OTHER ICY MATERIALS PRESENT IN THE ATMOSPHERE, GIVING IT A SIMILAR BLUE COLOUR AND THE TITLE OF AN "ICE GIANT". NEPTUNE ALSO HAS RINGS, BUT BEING EXTREMELY FAINT, THEY ARE NOT VISIBLE CLEARLY.

BEING THE FURTHEST PLANET FROM THE EARTH, IS IT EXTREMELY HARD TO REACH, AND TILL DATE, ONLY VOYAGER 2, A SPACE PROBE LAUNCHED BY NASA, HAS REACHED NEPTUNE.

DWARF PLANETS

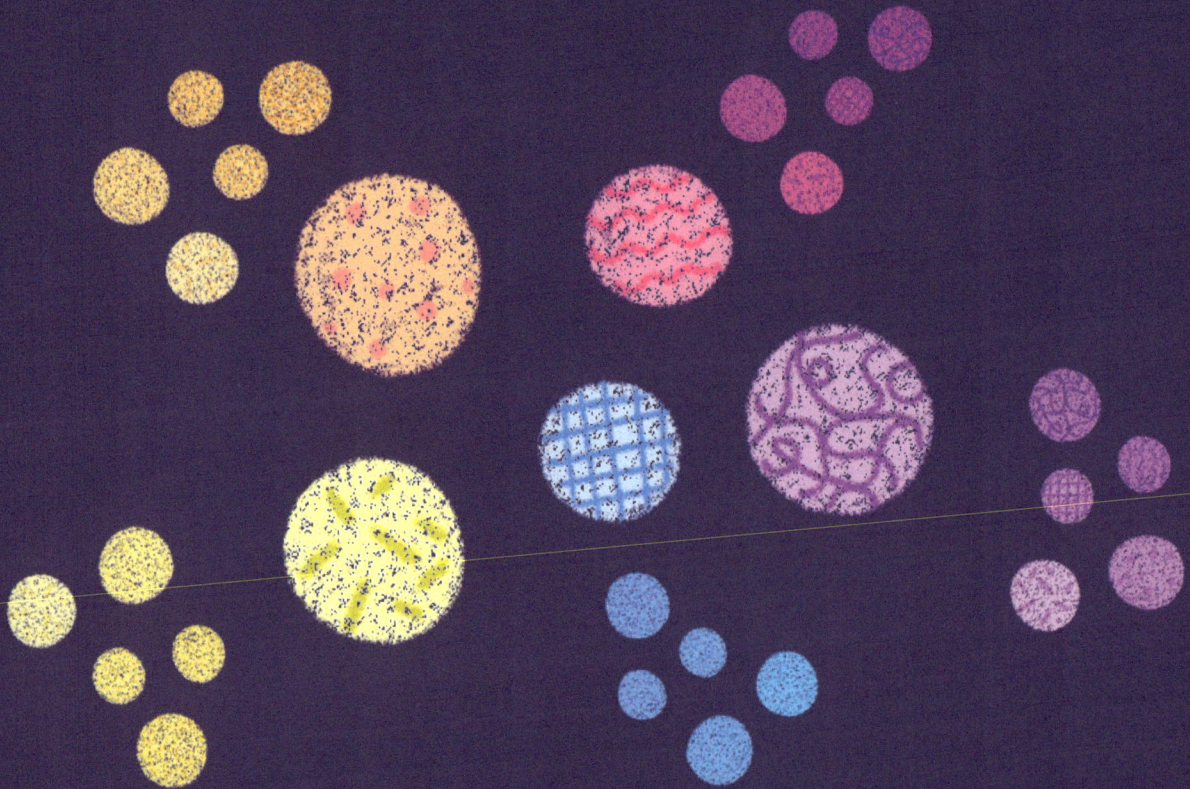

A BUD TO A FLOWER, A PLANT TO A TREE, A TOFFEE TO A CHOCOLATE, A SINGLE SCOOP TO A DOUBLE SCOOP AND A DWARF PLANET TO A PLANET. THESE ARE ALL THE SMALLER, TINIER VERSIONS OF THE LATTER.

DWARF PLANETS ARE THOSE CELESTIAL BODIES WHICH DO NOT FULFILL BOTH THE DEFINITIONS FOR PLANETARY SCIENCE SET BY THE INTERNATIONAL ASTRONOMICAL UNION (IAU):

1. IT IS MASSIVE ENOUGH TO BE IN HYDROSTATIC EQUILIBRIUM HAS CLEARED THE NEIGHBORHOOD AROUND ITS ORBIT
2. IT HAS CLEARED THE NEIGHBORHOOD AROUND ITS ORBIT.

WE HAVE DISCOVERED 5 DWARF PLANETS IN OUR SOLAR SYSTEM TILL NOW: PLUTO, CERIS, ERIS, MAKEMAKE AND HAUMEA. LET US NOW EXTENSIVELY TALK ABOUT THE MOST POPULAR OF THE LOT: PLUTO!

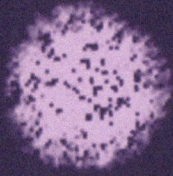

PLUTO

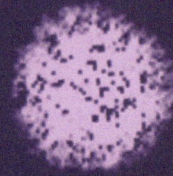

PLUTO, FORMERLY KNOWN AS THE 9TH AND THE SMALLEST PLANET OF THE SOLAR SYSTEM, BECAME THE SECOND EVER AND THE BIGGEST DWARF PLANET KNOWN TO US IN 2006. NAMED AFTER THE GOD OF THE UNDERWORLD, PLUTO HAS 5 MOONS. ITS LARGEST MOON, CHARON IS SO BIG THAT THE TWO CAN BE CONSIDERED A DOUBLE-BODY SYSTEM.

PLUTO WAS KICKED OUT OF THE CATEGORY OF PLANETS AFTER 76 YEARS OF ITS DISCOVERY.

ONE MAJOR REASON FOR PLUTO TO BE REMOVED FROM THE CATEGORY OF PLANETS WAS ITS SIZE. IT IS SO SMALL THAT ITS DIAMETER IS EQUAL TO THE WIDTH OF THE UNITED STATES!

PLUTO, PLACED IN THE KUIPER BELT, HAS ALSO BEEN GIVEN THE NICKNAME OF THE "KING OF THE KUIPER BELT", BEING DERIVED FROM THE FACT THAT IT IS THE LARGEST CELESTIAL BODY IN THAT REGION.

WITH AN EXTREMELY TILTED AND OVAL ORBIT, PLUTO'S ANGLE AND PATH OF ORBIT ARE QUITE DIFFERENT FROM THE REST.

THE LARGE DISTANCE FROM THE SUN IS CORRECTLY WITNESSED IN THE TIME OF ONE REVOLUTION BY THE PLANET, AS ONE YEAR ON PLUTO IS EQUAL TO 248 EARTH YEARS!

IT HAS A COLD, DARK ATMOSPHERE, WHERE THE TEMPERATURE DROPS TO -400 DEGREES FAHRENHEIT AND THE SUN IS JUST A SPECK OF GLITTER IN THE SKY.

● ● ● ● ● ● ● ● ● ● ● ● ● ● ● ●

BUT THIS DISTANCE WASN'T ENOUGH TO STOP US FROM KNOWING MORE ABOUT THIS INTERESTING OBJECT. ON THE 19TH OF JANUARY, 2006, NASA LAUNCHED A ROBOT SPACECRAFT UNDER THE PROJECT "NEW HORIZON" TO GAIN A NEW PERSPECTIVE OF THE BODIES AT THE VERY EDGE OF OUR SOLAR SYSTEM.

THE SPACECRAFT ARRIVED AT PLUTO IN JULY 2015 AND AIDED US IN THE STUDY TILL 2022.

● ● ● ● ● ● ● ● ● ● ● ● ● ● ●

SATELLITES

DIZZY, NAUSEOUS, VERTIGINOUS, FAINTING: THESE ARE THE FIRST WORDS I WOULD EXPECT TO HEAR FROM A SATELLITE UPON INITIATING A CONVERSATION. BUT, WHAT EVEN IS A SATELLITE? LET US SIT BACK AND ENJOY A STORYTELLING SESSION ABOUT AN OBJECT WHOSE ENTIRE WORLD REVOLVES AROUND SOMEONE ELSE, LITERALLY.

ONCE UPON A TIME, IN A LAND FAR FAR AWAY, EXACTLY 384,400 KM, A FAN-FAVOURITE DANCE PERFORMANCE, "DANCING WITH THE SATS".

IT WAS ATTENDED BY EVERY PLANET IN THE SOLAR SYSTEM EXCEPT MERCURY AND VENUS. THE PRODUCTION DISPLAYED THE SYNCHRONIZED MOVEMENTS OF THE DUO "THE SATELLITES", STARRING NATURAL AND ARTIFICIAL SATELLITES.

ACT 1: NATURAL SATELLITES (MOONS)
ACT 2: ARTIFICIAL SATELLITE (INTERNATIONAL SPACE STATION)

ACT 1 STARS OUR FIRST PERFORMERS: THE "NATURAL SATELLITES". THE GRACEFUL KATHAK DANCERS, TAKING "CHAKKARS" AROUND THEIR PARTNERS, GRACED THE STAGE. THESE DANCERS, MORE COMMONLY CALLED MOONS, ARE THE LOYAL COMPANIONS OF PLANETS, ADDING CHARM AND MYSTERY TO THEIR NIGHTLY ROUTINES.

JUPITER AND SATURN BOAST THEIR ENTOURAGE OF MOONS, EACH WITH ITS OWN UNIQUE PERSONALITY AND ORBITING STYLE. BUT THE REAL STAR OF ACT 1 IS THE GRAVITATIONAL FORCE THAT KEEPS THESE MOONS TWIRLING IN PERFECT HARMONY WITH THEIR PLANETARY PARTNERS.

AS ACT 1 CONCLUDES, A NEW SET OF PERFORMERS TAKES CENTER STAGE — THE ARTIFICIAL SATELLITES.

THESE MAN-MADE MARVELS ARE QUITE OBEDIENT, AS THEY PIROUETTE AROUND OUR EARTH IN THEIR ASSIGNED ORBITS, PERFORMING THE ASSIGNED TASKS TO THE BEST OF THEIR ABILITIES. FROM WEATHER FORECASTING TO G.P.S. NAVIGATION, THESE SATELLITES JUGGLE VARIOUS JOBS, MAKING OUR LIVES EASIER AND MORE CONNECTED.

THE AUDIENCE APPLAUDS THE TALENTED PERFORMERS.

WITH BOTH THE PERFORMING GROUPS ON STAGE, THE ULTIMATE SEQUENCE BEGINS.

AS THE MOONLIGHT SONATA (3RD MOVEMENT) BEGINS TO PLAY, THE SATELLITES STARTS. BUT THIS TIME, HARMONIZING WITH ONE ANOTHER, IT IS A SPECTACLE LIKE A GARDEN WITH BEES AND BUTTERFLIES. THE NATURAL SATELLITES, WITH THEIR NOBLE RADIANCE AND LUMINOSITY, GLOW AS THEY MOVE, FORMING STREAKS OF HALO AS THEY GLIDE OVER THE STAGE.

ON THE OTHER WING, THE ARTIFICIAL SATELLITES CARRY INFORMATION AND KNOWLEDGE GALORE, SIGNIFICANT CENTURIES WORTH OF HUMAN EFFORT AND PERSEVERANCE CARRIED THROUGH THE ATMOSPHERE.

AS THE PIANO NOW SWITCHES TO THE "FLIGHT OF THE BUMBLEBEE", THE TWO FACTIONS PREPARE FOR THE FINAL FORMATION, AND WITH THE WEIGHT OF THE LAST KEY, THEY CONCLUDE THEIR PHENOMENON DISPLAY OF DANCE, DRAMA AND BRILLIANT SYNERGY, PROPELLING TEARS AND COMPELLING THE AUDIENCE TO GIVE A STANDING OVATION, A SHOW OF SUCH MAGNIFICENCE THAT EVEN MERCURY AND VENUS REGRET MISSING.

STARS

CANARY, OXFORD BLUE, IVORY, TANGERINE, SCARLET: GLOWING, ALMOST NEON, LIKE A FLUORESCENT GREEN TENNIS BALL IN BRIGHT LIGHT, EXCEPT 445.26 BILLION TIMES LARGER.

STARS ARE BORN IN VAST CLOUDS OF GAS AND DUST KNOWN AS NEBULAE. LIKE A CHESSBOARD UPON THE DAWN OF A NEW GAME, WITH EACH PIECE—EVERY HYDROGEN ATOM—READY TO PLAY ITS PART, GEARS UP TO TOIL THROUGHOUT THE GAME, WITH ONE AIM TO CREATE A GLOWING ORB OF HEAT AND LIGHT.

GRAVITY IS THE GRANDMASTER HERE, PULLING THE GAS AND DUST TOGETHER, CAUSING THE CLOUD TO COLLAPSE AND HEAT UP. AS THE TEMPERATURE SOARS TO MILLIONS OF DEGREES, NUCLEAR FUSION IGNITES IN THE CORE, AND VOILA—A STAR IS BORN!

AT ITS CORE, A STAR IS LIKE A NUCLEAR REACTOR, PRIMARILY COMPOSED OF HYDROGEN AND HELIUM. HYDROGEN ATOMS FUSE TOGETHER TO FORM HELIUM, RELEASING A TREMENDOUS AMOUNT OF ENERGY. THIS FUSION IS WHAT MAKES STARS SHINE SO BRIGHTLY, MUCH LIKE A WELL-PLANNED STRATEGY ILLUMINATES THE PATH TO VICTORY IN CHESS.

JUST AS EACH CHESS PIECE HAS A SPECIFIC ROLE, THE ELEMENTS WITHIN A STAR PLAY CRUCIAL PARTS IN ITS LIFECYCLE.

STARS, LIKE CHESS PLAYERS, GO THROUGH DIFFERENT STAGES. THEIR JOURNEY DEPENDS ON THEIR INITIAL MASS, AND HERE'S HOW IT GENERALLY GOES:

- MAIN SEQUENCE: THIS IS THE STAR'S HEYDAY, AKIN TO THE MIDDLE GAME IN CHESS WHERE STRATEGIES ARE FULLY IN MOTION. FOR MOST OF ITS LIFE, A STAR IS IN THIS STABLE PHASE, FUSING HYDROGEN INTO HELIUM IN ITS CORE.

- RED GIANT/SUPERGIANT: WHEN A STAR EXHAUSTS ITS HYDROGEN FUEL, IT STARTS TO BURN HELIUM AND OTHER ELEMENTS. THE CORE CONTRACTS WHILE THE OUTER LAYERS EXPAND, AND THE STAR SWELLS INTO A RED GIANT OR SUPERGIANT. IT'S LIKE THE MID-GAME SHIFT IN CHESS, WHERE UNEXPECTED MOVES CAN CHANGE THE BOARD DRAMATICALLY.

- FINAL ACT: THE STAR'S FINAL STAGE DEPENDS ON ITS MASS.
  - LOW TO MEDIUM MASS STARS: THESE STARS SHED THEIR OUTER LAYERS, CREATING BEAUTIFUL PLANETARY NEBULAE, AND THE CORE BECOMES A WHITE DWARF. LIKE A STALEMATE SITUATION, WHERE THE GAME WINDS DOWN BUT THE CORE—LIKE A RESILIENT CHESS PLAYER—REMAINS.
  - HIGH MASS STARS: THESE STARS GO OUT WITH A BANG—A SUPERNOVA EXPLOSION. WHAT'S LEFT CAN BE EITHER A NEUTRON STAR OR, IF THE STAR WAS MASSIVE ENOUGH, A BLACK HOLE. TALK ABOUT AN EPIC CHECKMATE!

STARS COME IN VARIOUS TYPES, MUCH LIKE THE DIVERSE STYLES OF CHESS PLAYERS:

- RED DWARFS: THESE SMALL, COOL STARS BURN THEIR FUEL SLOWLY AND CAN LIVE FOR TRILLIONS OF YEARS. THEY'RE THE LONG-GAME STRATEGISTS OF THE STAR WORLD.
- YELLOW DWARFS: THESE STARS ARE MEDIUM-SIZED AND LIVE FOR BILLIONS OF YEARS. THEY'RE BALANCED AND STEADY PLAYERS.
- BLUE GIANTS: MASSIVE AND HOT, THEY BURN THROUGH THEIR FUEL QUICKLY, DYING YOUNG. THEY'RE THE AGGRESSIVE PLAYERS OF THE COSMIC CHESSBOARD.
- SUPERGIANTS: THESE ARE AMONG THE LARGEST AND MOST LUMINOUS STARS. THEIR SHORT LIVES END IN SPECTACULAR SUPERNOVAE, LIKE A DRAMATIC CHECKMATE IN A HIGH-STAKES GAME.

ASTEROIDS

8, 9, 4, SCRATCH, 2, ERASE. ANALOGOUS TO HOW MY ERASER ELIMINATES THE NUMBER 3 FROM THE SECOND BOX, THE ASTEROID MASS EXTINCTION WIPED DINOSAURS FROM THE FACE OF THE EARTH, 66 MILLION YEARS AGO. SIMILAR TO MY DAD DEFEATING ME AT RAPID-FIRE SUDOKU FOR THE FIRST TIME, AS ANTIHERO OF A ROLE AS ASTEROIDS MAY ABSORB POST MY DRAMATIC ANALOGY, THEY ARE WORTH ADMIRING!

STARS ARE THE FORGES OF THE UNIVERSE, CREATING THE ELEMENTS THAT MAKE UP EVERYTHING AROUND US. WHEN THEY DIE, THEY SCATTER THESE ELEMENTS INTO SPACE, SEEDING FUTURE GENERATIONS OF STARS AND PLANETS.

IN A WAY, WE ARE ALL MADE OF STARDUST—QUITE LITERALLY! IT'S LIKE THE ENDGAME IN CHESS, WHERE THE OUTCOMES OF PREVIOUS MOVES DETERMINE THE FINAL STATE OF THE BOARD.

SO, NEXT TIME YOU LOOK UP AT THE NIGHT SKY AFTER A CHESS MATCH, REMEMBER THAT EACH STAR HAS ITS OWN STORY, A TALE OF CREATION, STRUGGLE, AND TRANSFORMATION THAT SPANS BILLIONS OF YEARS.

BY STUDYING THEM, SCIENTISTS LEARN MORE ABOUT HOW PLANETS, INCLUDING EARTH, WERE FORMED. THEIR SURFACES ARE COVERED WITH SCARS AND CRATERS THAT TELL STORIES OF THEIR LONG JOURNEYS THROUGH SPACE, SURVIVING COLLISIONS AND COSMIC STORMS.

ASTEROIDS ALSO OFFER EXCITING POSSIBILITIES FOR FUTURE EXPLORATION. THEY COULD BE NEW PLACES FOR US TO VISIT OR MINE FOR RESOURCES. SOME ASTEROIDS ARE RICH IN METALS AND OTHER ELEMENTS THAT ARE RARE ON EARTH.

MOREOVER, ASTEROIDS HAVE A ROLE IN SHAPING OUR SOLAR SYSTEM. THEIR GRAVITATIONAL PULL INFLUENCES THE ORBITS OF PLANETS AND MOONS.

THEY MIGHT EVEN HELP US UNDERSTAND HOW LIFE BEGAN ON EARTH. SOME SCIENTISTS BELIEVE THAT ASTEROIDS COULD HAVE BROUGHT WATER AND ORGANIC MATERIALS TO OUR PLANET, SPARKING THE DEVELOPMENT OF LIFE.

SO, NEXT TIME YOU THINK ABOUT ASTEROIDS, DON'T JUST REMEMBER THEM AS THE TINY SUDOKU PRINT TO MY EYES OR DIGITAL SUDOKU SUBSCRIPTION FEE TO MY LIFE. THEY ARE FASCINATING ASTRAL ROCKS WITH A LOT TO TEACH US ABOUT OUR ORIGINS AND FUTURE ADVENTURES.

GOVERNED BY GRAVITATIONAL FORCES, THEY ARE CONTROLLED AS THE LINE DIVISIONS MANAGE AND SEPARATE EACH DIGIT IN THE 9X9 GRID.

EVERY MISTAKE LEADS TO REVELATIONS AND NEW SITUATIONS, AND MASS EXTINCTION, WHILE ELIMINATING THE DINOSAUR SPECIES, BOOSTS THE GROWTH OF FERNS.

SO, LET US CHERISH THESE ROCKY WANDERERS FOR THE VALUABLE KNOWLEDGE THEY BRING AND THE EXCITING OPPORTUNITIES THEY OFFER FOR EXPLORATION.

ASTEROID BELT

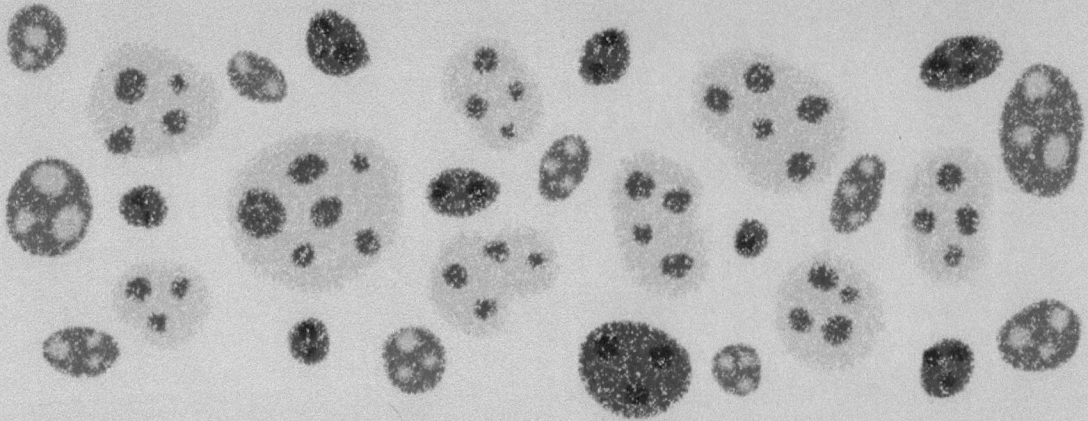

IN THE VAST EXPANSE BETWEEN MARS AND JUPITER LIES A CELESTIAL CHOIR, A BREATHTAKING COSMIC "KAWALI" KNOWN AS THE ASTEROID BELT.

AT THE DAWN OF OUR SOLAR SYSTEM, A GRANDIOSE CHORUS BEGAN TO UNFOLD. THE GRAVITATIONAL FORCES OF PLANETS SCULPTED THE VERY LAYERS OF THE COSMOS. A MULTITUDE OF ROCKY REMNANTS OF PLANETARY FORMATION FOUND THEIR PLACE IN THE SUNLIT ROWS BETWEEN MARS AND JUPITER.

THESE FRAGMENTS, RANGING IN SIZE FROM TINY PEBBLES TO BOULDERS, BEGAN THEIR MUSICAL EXHIBITIONS AROUND THE SUN, CAUGHT IN THE GENTLE GRAVITATIONAL PULL OF THE NEIGHBOURING PLANETS. THE STAGE WAS SET, AND THE ASTEROID BELT EMERGED AS A CELESTIAL "KAWALI".

AS THE COSMIC ENSEMBLE UNFOLDS, THE ASTEROID BELT REVEALS ITS DIVERSITY AS A STAGE SET WITH SINGERS SINGING BRIDGES AND MUSICIANS READY WITH THEIR HARPS AND HARMONIUMS.

THESE ASTEROIDS COME IN ALL SHAPES AND SIZES, EACH WITH ITS UNIQUE APPEARANCE AND PERSONALITY. SOME ARE METALLIC, GLISTENING LIKE PRECIOUS JEWELS IN THE DAZZLING LIGHT, WHILE OTHERS ARE STONY AND RUGGED, BEARING THE SCARS OF AEONS OF COSMIC COLLISIONS.

THE DIVERSITY OF THE ASTEROID BELT GOES BEYOND ITS PHYSICAL ATTRIBUTES, EXTENDING TO THEIR CORE COMPOSITIONS.

SOME ASTEROIDS ARE CLASSIFIED AS C-TYPES, RICH IN CARBON COMPOUNDS, WHILE OTHERS ARE S-TYPES, COMPOSED OF SILICATE ROCK. THE HARMONY OF ELEMENTS CREATES A MASTERPIECE THAT BECKONS ASTRONOMERS AND STARGAZERS ALIKE TO GAZE IN WONDER.

THE ASTEROIDS, THESE ANCIENT RELICS OF OUR SOLAR SYSTEM'S FORMATION, HOLD THE KEYS TO UNLOCKING THE MYSTERIES OF OUR ETHEREAL PAST. HUMAN EXPLORATION AND ROBOTIC MISSIONS HAVE SET THEIR SIGHTS ON THIS MELODIOUS GROUP, EAGER TO UNRAVEL THE SECRETS HELD BY THE ASTEROIDS.

THE GRAND FINALE OF THIS CAROL IS YET TO BE WRITTEN. THE ASTEROID BELT CONTINUES TO ENCHANT US WITH ITS DIVERSITY, GRACE, AND SCIENTIFIC INTRIGUE.

COMETS

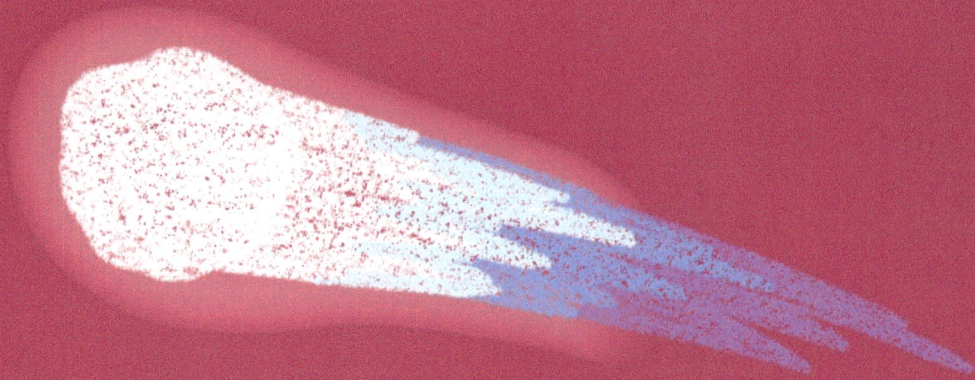

Throughout history, comets have been seen as harbingers of change. Ancient civilizations sometimes interpreted them as omens, believing they foretold significant events. For me, they are reminders of those chilly nights in Ladakh when the universe felt a little closer.

These are made of ice, dust, and rock, making them similar to dirty snowballs. When they get close to the sun, they start to heat up, the ice melts and turns into gas, creating a glowing head and a long, shining tail.

THESE PILGRIMS FOLLOW ELONGATED, OVAL-SHAPED PATHS IN THEIR ORBITS AROUND THE SUN. WHEN COMETS PASS CLOSE TO EARTH, THEY CAN BECOME VISIBLE TO THE NAKED EYE, APPEARING AS A BRIGHT NUCLEUS SURROUNDED BY A DIFFUSE, ETHEREAL TAIL THAT GLOWS WITH REFLECTED SUNLIGHT.

COMETS ARE NAMED FOR THEIR DISCOVERERS OR THE SPACE MISSIONS THAT IDENTIFY THEM, SUCH AS "HALLEY'S COMET" OR "COMET NEOWISE". SOME COMETS ARE PERIODIC, RETURNING TO OUR VICINITY AT PREDICTABLE INTERVALS, WHILE OTHERS ARE NON-PERIODIC, MAKING A FLEETING APPEARANCE BEFORE HEADING BACK INTO THE DEPTHS OF SPACE.

THESE CELESTIAL WANDERERS HAVE FASCINATED HUMANS FOR CENTURIES, INSPIRING AWE AND CURIOSITY ABOUT OUR SOLAR SYSTEM'S MYSTERIES.

OBSERVING COMETS IS A REMINDER THAT OUR SOLAR SYSTEM IS A DYNAMIC PLACE, WHERE ICY VISITORS FROM THE OUTER REACHES BRING WONDER AND BEAUTY TO OUR NIGHT SKIES.

SO, WHEN YOU SPOT A COMET WITH ITS GLOWING TAIL STRETCHING ACROSS THE DARKNESS, TAKE A MOMENT TO APPRECIATE THIS ANCIENT TRAVELER FROM THE DISTANT CORNERS OF OUR COSMIC NEIGHBORHOOD.

THE UNIVERSE

BEHOLD, FOR NOW, WE ARE GOING TO DELVE INTO THE MOSAIC OF OUR EXISTENCE, MADE PIECE BY PIECE BY THE TILES OF THE SPACE-TIME FABRIC — THE UNIVERSE.

OUR UNIVERSE IS WHERE EVERY OLD THING DIES AND EVERY NEW ONE TAKES BIRTH, THE PLACE WHERE SIRIUS BLACK DIED TRAGICALLY AND SIRIUS THE STAR SHINES EVER SO BRIGHT, THE PLACE WHERE CAPTAIN AHAB MAPS OCEANS AND THE PLACE WHERE YOU READ THIS BOOK WHILE SECRETLY WANTING TO SCUBA DIVE IN THE ANDAMANS, AND THE PLACE WHERE I TYPE THESE WORDS OUT LOOKING FORWARD TO THE SCARIEST YEAR OF MY LIFE.

ALL OF THIS, THE ATOMS AND THE BOULDERS, THE THREADS AND THE SARIS, THE BRICKS AND THE SKYSCRAPERS, ALL COME TOGETHER IN THIS CONGLOMERATE OF A MASTERPIECE.

ON A LARGER SCALE, GALAXIES, COLOSSAL CONCOCTIONS OF STARS, GAS, AND COSMIC DUST FRAME THE MAIN NARRATIVE OF THIS FABLE.

TRAVELING TRILLIONS OF LIGHT-YEARS, LIGHT UNCOVERS TRUTHS ABOUT THE UNIVERSE LIKE NO ONE ELSE. FROM INFORMATION ABOUT EXTRA-TERRESTRIAL PLANETS, UNKNOWN GALAXIES, DARK MATTER AND BLACK HOLES TO DETAILS ABOUT THE EARLY UNIVERSE AND THE BIG BANG, WHATEVER THE BEAMS OF LIGHT TOUCH IS PRINTED AS OUR LATEST DISCOVERY IN THE FIELD OF SPACE SCIENCE.

THE UNIVERSE IS DICTATED BY THE FORCE OF GRAVITY, AS IT MOULDS AND DISTORTS THE SPACE-TIME FABRIC, THE 4-DIMENSIONAL WEB WHICH MAKES UP OUR UNIVERSE. THIS FORCE DECIDES THE DIRECTION OF THE FALLING APPLE, THE MOVEMENT OF PLANETS AROUND A STAR AND THE FLOW OF LIGHT.

IT IS THE ULTIMATE MASTERMIND IN THIS SCI-FI THRILLER NOVEL.

As you contemplate the universe, envision it as a notebook with work done on every subject, along with tic-tac-toe hatches and professional signature practices scattered at the end. After all, it truly contains everything—both seen and unseen—that surrounds us.

It is the one drawer that we stuff everything into, the stationary pouch which has every possible colour of highlighter. An infinitely increasing universe is hard to imagine, like the thought of one moving to college in 8 months, but it is true. There are various theories about the universe, and hence, make it our responsibility to learn more about this enigmatic home of ours and discover what it has in store.

GALAXIES

DELHI'S CONNAUGHT PLACE, BENARAS' MANDIRS, PORT BLAIR'S SEAS AND LUCKNOW'S "TEHZEEB" ARE WHAT DIFFERENTIATE THEM FROM ONE ANOTHER, EACH WITH AN ECCENTRIC ORIGIN AND UNEQUALLED HERITAGE.

SIMILARLY, GALAXIES AREN'T JUST RANDOM GROUPS OF STARS; THEY ARE LIKE BUSY CITIES, EACH WITH ITS STYLE. SPIRALS TWIST BEAUTIFULLY LIKE COSMIC PINWHEELS, ELLIPTICALS ARE SMOOTH AND ELEGANT, AND IRREGULARS ARE WILD AND UNTAMED.

INSIDE THESE STELLAR CITIES, STARS ARE THE RESIDENTS, LIGHTING UP THE WINDOWS. THE GRAVITY OF THESE STARS CREATES PATHS WHERE PLANETS, MOONS, AND ASTEROIDS MOVE IN STRUCTURED STREETS.

THE REAL MAGIC IS IN THE UNSEEN FORCES THAT HOLD THESE GALAXIES TOGETHER. DARK MATTER, A MYSTERIOUS AND INVISIBLE SUBSTANCE, ACTS AS THE FRAMEWORK THAT SHAPES AND SUPPORTS GALAXIES.

IT'S LIKE A HIDDEN BLUEPRINT GUIDING THE CONSTRUCTION OF THESE GIANT STRUCTURES. GALAXIES, MUCH LIKE THE SHAPE OF CITIES, ARE UNIQUE AND DIVERSE. SOME HAVE HUNGRY BLACK HOLES AT THEIR CENTRES, DEVOURING NEARBY MATTER. OTHERS ARE NURSERIES WHERE NEW STARS ARE BORN IN BRIGHT BURSTS OF LIGHT.

WHEN WE LOOK UP AT THE NIGHT SKY, WE ARE PEERING INTO THE HEART OF THESE GALACTIC CITIES, SEEING THE ANCIENT LIGHT OF DISTANT STARS AND THE PROMISE OF UNKNOWN WORLDS.

GALAXIES ARE NOT JUST CELESTIAL OBJECTS.

JUST LIKE THE MONUMENTS IN CITIES NARRATE THE TALES OF THEIR HERITAGE, SIMILARLY THE GALAXIES ARE THE STORYTELLERS OF THE UNIVERSE, TELLING FABLES OF COSMIC COLLISIONS, STAR BIRTHS, AND THE PASSAGE OF TIME.

THE MILKY WAY, THE GALAXY IN WHICH WE RESIDE, IS APPROXIMATELY 13.61 BILLION YEARS OLD, AND OUR NEIGHBOR, THE ANDROMEDA GALAXY IS CLOSE TO 10.01 BILLION YEARS INTO ITS EXISTENCE.

SO, AS THE CITIES SING A SONG ABOUT DELHI'S VIBRANT CHAAT, BENARAS' SACRED GHATS, LUCKNOW'S ELEGANT NAWABS AND PORT BLAIR'S EVERGREEN SPARK, LET US LOOK AT THE CELESTIAL CITIES OF OUR UNIVERSE AND LISTEN TO THE CHRONICLES THEY ARE EAGER TO BROADCAST.

BLACK HOLES

BLUE. DEEP, DARK, DAMP. THE DEPTH OF THE OCEAN CAN BE CHARACTERIZED BY THE SHADES OF BLUE. AS THE WEIGHT OF THE OXYGEN CYLINDER ON YOUR SHOULDERS REDUCES AND THE DISPERSED RAYS OF SUNLIGHT PAT YOUR SKIN, ONE FEELS AT EASE. SURROUNDED BY VIVID BOUQUETS OF BREATHING ROCKS AND SCHOOLS OF SWIMMING PROFESSIONALS, ONE FEELS AT HOME. I FEEL AT HOME. NOT JUST BECAUSE I HAVE AN INSANE AMOUNT OF LOVE FOR THE ART OF SCUBA DIVING, BUT ALSO BECAUSE CREATING WATER RINGS USING MY HANDS IS MY NEW HOBBY.

BLACK HOLES ARE REGIONS IN SPACE WHERE GRAVITY IS SO INTENSE THAT NOTHING, NOT EVEN LIGHT, CAN ESCAPE THEIR GRASP.

THEY FORM WHEN MASSIVE STARS COLLAPSE UNDER THEIR OWN GRAVITY AT THE END OF THEIR LIFE CYCLES. IT'S AKIN TO THE FEELING YOU GET WHEN YOU DIVE INTO THE DEEP BLUE SEA, WHERE THE PRESSURE INCREASES AND LIGHT STARTS TO FADE, LEAVING ONLY THE UNKNOWN AHEAD.

THE EVENT HORIZON IS THE "POINT OF NO RETURN." ANYTHING THAT CROSSES THIS BOUNDARY IS INEVITABLY PULLED INTO THE BLACK HOLE. LIKE THE EDGE OF A DEEP UNDERWATER TRENCH, ONCE YOU GO BEYOND, THE WAY BACK BECOMES IMPOSSIBLE.

SINGULARITY LIES AT THE VERY CENTER, A POINT WHERE DENSITY BECOMES INFINITE AND THE LAWS OF PHYSICS AS WE KNOW THEM CEASE TO APPLY.

IT'S EQUIVALENT TO THE DEEPEST PART OF THE OCEAN, A PLACE SO EXTREME AND MYSTERIOUS THAT IT DEFIES OUR CURRENT UNDERSTANDING.

BLACK HOLES, MUCH LIKE MARINE CREATURES, COME IN DIFFERENT TYPES AND SIZES:

• STELLAR BLACK HOLES FORM FROM THE REMNANTS OF MASSIVE STARS AND ARE SIMILAR TO THE COMMON BUT FASCINATING MARINE LIFE YOU ENCOUNTER ON MOST DIVES. THEY'RE ABUNDANT AND SCATTERED THROUGHOUT THE GALAXY.

- SUPERMASSIVE BLACK HOLES ARE FOUND AT THE CENTERS OF GALAXIES. THESE GIANTS HAVE MASSES EQUIVALENT TO MILLIONS AND BILLIONS OF SUNS. ENCOUNTERING ONE WOULD BE LIKE STUMBLING UPON A MASSIVE UNDERWATER CANYON TEEMING WITH EXTRAORDINARY CREATURES.

- PRIMORDIAL BLACK HOLES ARE LESS UNDERSTOOD AND MUCH LIKE THE ELUSIVE, RARELY SEEN CREATURES OF THE DEEP SEA, THEY SPARK CURIOSITY AND WONDER AMONG SCIENTISTS.

BLACK HOLES PLAY A CRUCIAL ROLE IN THE COSMOS. THEIR IMMENSE GRAVITATIONAL PULL CAN AFFECT THE ORBITS OF NEARBY STARS AND EVEN INFLUENCE THE GROWTH OF GALAXIES. JUST AS OCEAN CURRENTS SHAPE MARINE ECOSYSTEMS, BLACK HOLES SCULPT THE STRUCTURE OF THE UNIVERSE.

MY PASSION FOR SCUBA DIVING HAS GIVEN ME A UNIQUE PERSPECTIVE ON THE ALLURE AND MYSTERY OF BLACK HOLES. DIVING INTO THE DEPTHS OF THE OCEAN AND BEING SURROUNDED BY THE UNKNOWN MIRRORS THE JOURNEY SCIENTISTS EMBARK ON WHEN STUDYING BLACK HOLES. BOTH REALMS—THE DEEP SEA AND THE COSMIC ABYSS—REMIND US OF HOW MUCH THERE IS YET TO EXPLORE AND UNDERSTAND. THEY INVOKE A SENSE OF WONDER AND A DESIRE TO KEEP PUSHING THE BOUNDARIES OF OUR KNOWLEDGE.

SO, WHETHER IT'S THE THRILL OF DESCENDING INTO THE OCEAN'S DEPTHS OR THE QUEST TO UNRAVEL THE SECRETS OF BLACK HOLES, THE ADVENTURE LIES IN EMBRACING THE UNKNOWN AND MARVELLING AT THE EXTRAORDINARY WONDERS OF OUR UNIVERSE.

NEBULAE

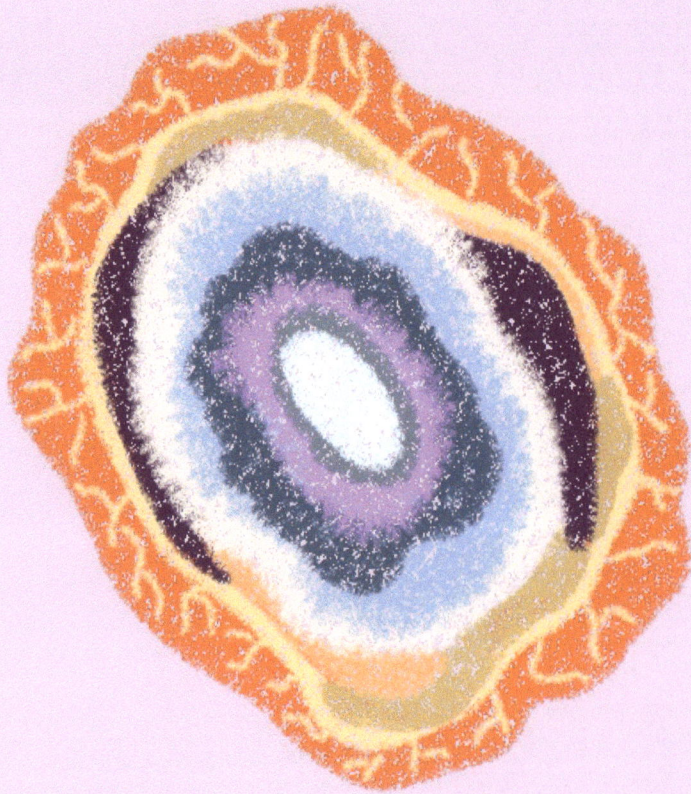

STANDING IN FRONT OF MY CLASS, NERVOUSLY CLUTCHING MY NOTES, READY TO TELL THEM ABOUT ONE OF THE MOST BRILLIANT PHENOMENA IN THE UNIVERSE: NEBULAE. THESE VAST, COLOURFUL CLOUDS OF GAS AND DUST WERE MY COSMIC MUSE, AND I LOVED SHARING THEIR STORIES WITH ANYONE WHO WOULD LISTEN.

NEBULAE ARE THE UNIVERSE'S GRANDEST ART INSTALLATIONS. THEY ARE BORN IN THE AFTERMATH OF STELLAR DEATH OR THE CHAOS OF STAR FORMATION, EACH ONE A TESTAMENT TO THE INCREDIBLE PROCESSES THAT SHAPE OUR COSMOS.

THESE ARE SEVERAL TYPES OF NEBULAE BASED ON THEIR RELATIONSHIP WITH LIGHT:

- EMISSION NEBULAE ARE THE SHOW-OFFS OF THE NEBULA FAMILY. EMISSION NEBULAE GLOW BRIGHTLY BECAUSE THEIR GASES ARE IONIZED BY THE HIGH-ENERGY RADIATION FROM NEARBY YOUNG STARS. STANDING ON STAGE UNDER THE BRIGHT LIGHTS, FEELING THE ENERGY AND EXCITEMENT OF THE AUDIENCE—THAT'S THE KIND OF VIBRANT PRESENCE THESE NEBULAE HAVE.

- REFLECTION NEBULAE REFLECT THE LIGHT OF NEARBY STARS, MUCH LIKE A PUBLIC SPEAKER WHO CHANNELS THE ENERGY OF THEIR AUDIENCE. THEIR BEAUTY IS SUBTLER, OFTEN APPEARING AS SOFT, BLUE-TINTED CLOUDS.

- DARK NEBULAE ARE MYSTERIOUS AND BROODING. THEY ARE DENSE CLOUDS OF GAS AND DUST THAT BLOCK THE LIGHT FROM STARS BEHIND THEM. THEY REMIND ME OF THOSE MOMENTS BEFORE A BIG SPEECH WHEN EVERYTHING FEELS UNCERTAIN AND THE SPOTLIGHT HASN'T YET REVEALED YOUR STORY.

- JUST AS PREPARATION FOR A SPEECH, FROM GATHERING THOUGHTS, ORGANIZING IDEAS, TO REHEARSING DELIVERY, FORMS THE BASE OF THE PERFORMANCE, SIMILARLY, NEBULAE FORM THE BASE IN THE COSMIC CYCLE OF BIRTH, DEATH, AND REBIRTH OF STARS.

STARS BORN WITHIN THESE NURSERIES LIVE OUT THEIR BRILLIANT LIVES, AND EVENTUALLY CONTRIBUTE TO THE FORMATION OF NEW NEBULAE WHEN THEY DIE.

AS I REACHED MY PEAK FORM OF EXPLANATION, STROLLING IN FRONT OF THE BLACKBOARD, REGULARLY ADJUSTING MY TONE, I WAS IMMERSED IN THE BEAUTY OF THIS ETHEREAL PHENOMENA.

I FELT A CONNECTION TO THIS SPECTACLE. IT TAUGHT ME THAT JUST LIKE IN PUBLIC SPEAKING, THERE'S BEAUTY IN BOTH THE BOLD AND THE SUBTLE, THE LOUD AND THE SOFT, THE BRIGHT AND THE DARK. EVERY NEBULA, LIKE EVERY SPEECH, HAS ITS UNIQUE STORY AND PURPOSE, UNPARALLELED, WITH A JOURNEY AND GOAL TO FULFILL.

SO NEXT TIME YOU LOOK UP AT THE NIGHT SKY AND SEE A NEBULA, REMEMBER THAT THEY'RE NOT JUST BEAUTIFUL CLOUDS OF GAS AND DUST— THEY'RE SPEAKERS, JUST LIKE US. THEY REMIND US OF THE POWER OF TRANSFORMATION, THE IMPORTANCE OF EVERY MOMENT, AND THE ENDLESS POSSIBILITIES THAT COME FROM SHARING OUR PASSIONS WITH THE WORLD.

THE KUIPER BELT

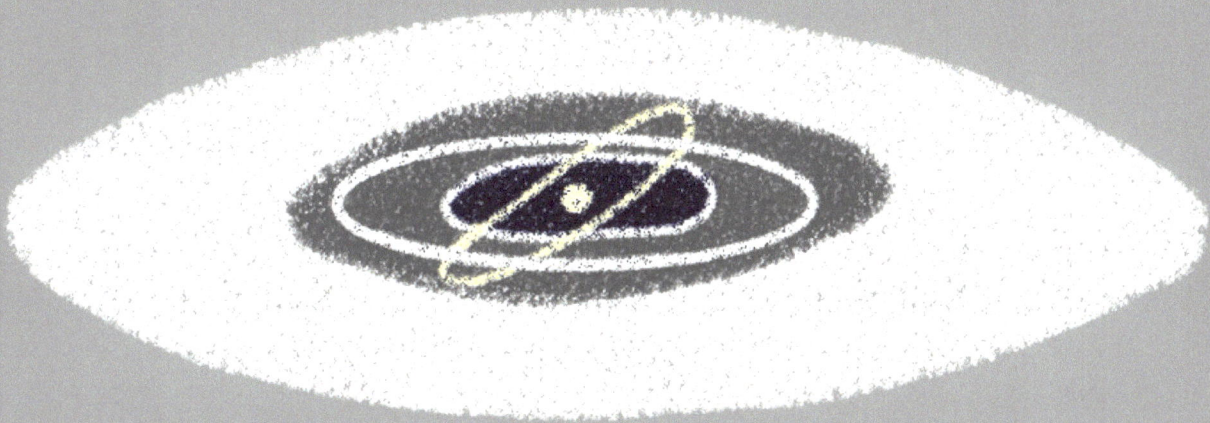

IN OUR VAST AND PERPLEXING SOLAR SYSTEM, A
PECULIAR NEIGHBORHOOD EXISTS BEYOND THE FAMILIAR
PLANETS — A PLACE AS CHAOTIC AS THE TITLE OF
THIS BOOK — KNOWN AS THE KUIPER BELT. THIS
HEAVENLY JUNKYARD, WHERE ICY ODDITIES AND
DWARF PLANETS RESIDE, ADDS A QUIRKY TWIST TO
OUR PLANETARY COMMUNITY.

THE KUIPER BELT, A DISTANT REGION STRETCHING FROM THE ORBIT OF NEPTUNE TO ABOUT 55 ASTRONOMICAL UNITS FROM THE SUN, IS AKIN TO A GALACTIC TREASURE TROVE OF REMNANTS FROM THE EARLY DAYS OF OUR SOLAR SYSTEM.

IT'S A CELESTIAL ATTIC, PRESERVING CLUES ABOUT THE FORMATION AND EVOLUTION OF THE PLANETS, AND A FASCINATING ARENA WHERE THE ECCENTRICITIES OF SPACE PLAY OUT.

THIS REGION, A BUSTLING CITY WHERE PLUTO — ONCE THE NINTH PLANET — DONS THE HAT OF THE MAYOR. THE DWARF PLANETS ERIS, HAUMEA, MAKEMAKE, AND A MYRIAD OF OTHER ICY BODIES ARE THE MINISTERS, EACH WITH THEIR OWN UNIQUE STORY ETCHED IN FROZEN WATER AND EXOTIC ICE.

THE MAIN ATTRACTION OF THE KUIPER BELT IS ITS CITIZENS — THE KUIPER BELT OBJECTS (KBOS). THESE ARE CELESTIAL NOMADS, REMNANTS FROM THE SOLAR SYSTEM'S EARLY DAYS, FROZEN IN TIME AND SPACE. THEY RANGE FROM TINY, COMET-SIZED BODIES TO THE MIGHTY DWARF PLANETS.

THE KUIPER BELT OBJECTS ARE NOT MERE SPACE DEBRIS; THEY ARE TIME CAPSULES HOLDING SECRETS ABOUT THE CONDITIONS THAT PREVAILED IN THE EARLY DAYS OF OUR SOLAR SYSTEM.

AS RESEARCHERS EXPLORE THIS DISTANT REGION, THEY UNRAVEL THE MYSTERIES OF THE SOLAR SYSTEM'S YOUTH, PIECING TOGETHER THE PUZZLE OF HOW PLANETS, MOONS, AND ASTEROIDS CAME TO BE.

COMETS PLAY THE ROLE OF THE CHILDREN OF THIS TOWN, ZIPPING AROUND IN THE KUIPER BELT PLAYGROUND, LEAVING TRAILS OF STARDUST IN THEIR WAKE.

THESE COMETS, INHABITANTS OF THE KUIPER BELT, EMBARK ON JOURNEYS THAT BRING THEM CLOSE TO THE INNER SOLAR SYSTEM, CREATING DAZZLING DISPLAYS FOR THOSE LUCKY ENOUGH TO WITNESS THEIR FLEETING BRILLIANCE.

AS OUR COSMIC JOURNEY THROUGH THE KUIPER BELT COMES TO AN END, WE'RE LEFT WITH A SENSE OF AWE AND WONDER. THIS DISTANT REALM, OFTEN OVERLOOKED IN THE HELIOCENTRIC SYSTEM'S NARRATIVE, IS A TESTAMENT TO THE RICH DIVERSITY AND UNPREDICTABILITY THAT DEFINES OUR SURROUNDINGS.

THE KUIPER BELT, WITH ITS ICY ODDITIES AND DWARF PLANET CHARACTERS, INVITES US TO EMBRACE OUR BOUNDARIES AND TO CONTINUE PUSHING THEM, EXPLORING THE MYSTERIES THAT LIE BEYOND THE FAMILIAR PLANETS, IN THE VAST AND CAPTIVATING EXPANSE OF SPACE.

ACKNOWLEDGEMENTS

WRITING BEYOND THE KUIPER BELT HAS BEEN AN ABSOLUTE WHIRLWIND, AND I OWE IMMENSE GRATITUDE TO THE PEOPLE WHO MADE THIS PROJECT POSSIBLE.

FIRST AND FOREMOST, I WOULD LIKE TO THANK MY FAMILY FOR THEIR UNWAVERING SUPPORT AND BELIEF IN ME. THEIR CONSTANT ENCOURAGEMENT FUELED MY PASSION FOR SCIENCE AND STORYTELLING.

I AM DEEPLY GRATEFUL TO MY TEACHERS, WHOSE GUIDANCE AND PASSION FOR LEARNING HAVE INSPIRED ME TO EXPLORE THE UNIVERSE FAR BEYOND THE CONFINES OF TEXTBOOKS. YOUR MENTORSHIP HAS BEEN INSTRUMENTAL IN SHAPING MY ACADEMIC JOURNEY, AND YOUR ENCOURAGEMENT HAS FUELED MY CURIOSITY AND LOVE FOR DISCOVERY.

TO MY FRIENDS AND PEERS, THANK YOU FOR YOUR ENTHUSIASM AND PATIENCE AS I RAMBLED ON ABOUT SPACE AND GRAVITATIONAL LENSING. YOU KEPT ME GROUNDED WHILE I REACHED FOR THE STARS.

LASTLY, TO ALL THE YOUNG READERS AND ASPIRING SCIENTISTS OUT THERE—THIS BOOK IS FOR YOU. MAY YOUR CURIOSITY ABOUT THE UNIVERSE NEVER FADE, AND MAY YOU ALWAYS LOOK BEYOND THE KNOWN, TOWARDS THE WONDERS THAT LIE JUST BEYOND THE KUIPER BELT.

Milton Keynes UK
Ingram Content Group UK Ltd.
UKHW050206231124
451130UK00023B/118